Funtastic Math

Multiplication and Division

by Joyce Mallery

S C H O L A S T I C
PROFESSIONAL**B**OOKS

New York ✳ Toronto ✳ London ✳ Auckland ✳ Sydney

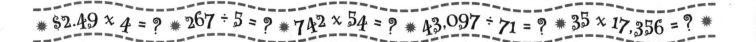

Edited by Sarah Glasscock

Cover design by Jaime Lucero and Vincent Ceci

Interior design by Ellen Matlach Hassell
for Boultinghouse & Boultinghouse, Inc.

Interior illustrations by Ellen Joy Sasaki and Manuel Rivera

ISBN 0-590-37366-8

Contents

(continued on the next page)

❋ Activity includes a student reproducible.

✳ Activity includes a student reproducible.

Introduction

With this book of activities, part of a six-book mathematics series, we hope to make teaching and understanding multiplication and division fun, creative, and exciting.

An Overview of the Book

Table of Contents

The table of contents features the activity names and page numbers, as well as stars to mark student reproducibles. Activities are categorized by multiplication or division topic, so you may use the table of contents as a scope and sequence.

Teaching Pages

Everything you need to know is on the teaching page, but you also have the option of tailoring the activities to meet students' individual needs and to address the wide variety of skills displayed in your classroom.

Learning Logo

A logo indicating the multiplication or division topic being discussed appears at the top of the page. The logo is correlated to the topics in the table of contents. This will enable you to key the activities to your mathematics curriculum quickly and easily.

Learning Objective

The objective clearly states the primary aim of the activity.

Grouping

This states whether the whole class, individual students, pairs, or cooperative groups should perform the task. If an activity lends itself to more than one grouping, the choices are indicated. Again, if you feel that another grouping is more appropriate to your classroom, feel free to alter the activity accordingly.

Materials

To cut your preparation time, all materials necessary for the main activity (including student reproducible) and its extension are listed. Most of the materials are probably already in your classroom. If an activity has a student reproducible with it, the page number of the reproducible is listed here.

Advance Preparation

A few activities require some minimal advance preparation on your part. All the directions you need are given here. You may also let students take over some or all of the preparation.

Directions

The directions usually begin with suggestions on how to introduce or review the multiplication or division topic, including any terms and/or formulas. Step-by-step details on how to do the activity follow. When pertinent, specific strategies that might help students in solving problems are suggested.

Taking It Farther

This section on the teaching page offers suggestions on how you can extend and enrich the activity. Students who require extra help and those who need a challenge will both benefit when you move the activity to a different level.

Assessing Skills

The key questions and/or common errors pointed out in this section will help alert you to students' progress. (In fact, you may want to jot down more questions on the page.) Use the information you gather about students here in conjunction with the teacher assessment form that appears on page 64 of the book.

Answers

When answers are called for, they appear at the bottom of the teaching page.

Student Reproducibles

About one-third of the activities have a companion student reproducible page for you to duplicate and distribute. These activities are marked with a star in the table of contents.

Do I Have Problems!

These pages are filled with fun and challenging Problems of the Day that you may write on the board or post on the bulletin board. The answers appear in brackets at the end of each problem.

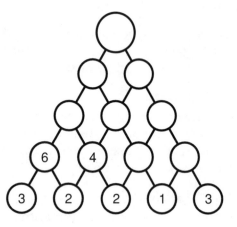

Assessment

Student Self-Evaluation Form

At the end of the activity, hand out these forms for students to complete. Emphasize that their responses are for themselves as well as you. Evaluating their own performances will help students clarify their thinking and understand more about their reasoning.

Teacher Assessment Form and Scoring Rubric

The sign of a student's success with an activity is more than a correct answer. As the NCTM stresses, problem solving, communication, reasoning, and connections are equally important in the mathematical process. How a student arrives at the answer—the strategies she or he uses or discards, for instance—can be as important as the answer itself. This assessment form and scoring rubric will help you determine the full range of students' mastery of skills.

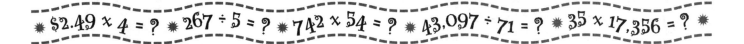
⊚→National Council of Teachers of Mathematics Standards

The activities in this book, and the rest of the series, have been written with the National Council of Teachers of Mathematics (NCTM) Standards in mind. The first four standards—Mathematics as Problem Solving, Mathematics as Communication, Mathematics as Reasoning, and Mathematical Connections—form the philosophical underpinning of the activities.

Standard 1: Mathematics as Problem Solving
The open-ended structure of the activities, and their extension, builds and strengthens students' problem-solving skills.

Standard 2: Mathematics as Communication
Class discussion at the beginning and ending of the activities is an integral part of these activities.

Additionally, communication is fostered when students work in pairs or cooperative groups and when individuals share and compare work.

Standard 3: Mathematics as Reasoning
Communicating their processes in working these activities gives students the opportunity to understand and appreciate their own thinking.

Standard 4: Mathematical Connections
A variety of situations has been incorporated into the activities to give students a broad base on which to apply mathematics. Topics range from real-life experiences (historical and contemporary) to the whimsical and fantastic, so students can expand their mathematical thinking to include other subject areas.

- -

More specifically, the activities in this book address the following NCTM Standards.

NCTM Standards Grades K–4:

Standard 5: Estimation
* Explore estimation strategies.
* Recognize when an estimate is appropriate.
* Determine the reasonableness of results.
* Apply estimation in working with quantities, measurement, computation, and problem solving.

Standard 6: Number Sense and Numeration
* Construct number meanings through real-world experiences and the use of physical materials.
* Understand our numeration system by relating counting, grouping, and place-value concepts.
* Develop number sense.
* Interpret the multiple uses of numbers encountered in the real world.

Standard 7: Concepts of Whole Number Operations
* Develop meaning for the operations by modeling and discussing a rich variety of problem situations.
* Relate the mathematical language and symbolism of operations to problem situations and informal language.
* Recognize that a wide variety of problem structures can be represented by a single operation.
* Develop operation sense.

Standard 8: Whole Number Computation
* Model, explain, and develop reasonable proficiency with basic facts and algorithms.
* Use a variety of mental computation and estimation techniques.
* Use calculators in appropriate computational situations.

* Select and use computation techniques appropriate to specific problems and determine whether the results are reasonable.

NCTM Standards Grades 5–8:

Standard 5: Number and Number Relationships
* Understand, represent, and use numbers in a variety of equivalent forms (integer, fraction, decimal, percent, exponential, and scientific notation) in real-world and mathematical problem situations.
* Develop number sense for whole numbers, fractions, decimals, integers, and rational numbers.

Standard 6: Number Systems and Number Theory
* Develop and order relations for whole numbers, fractions, decimals, integers, and rational numbers.
* Understand how the basic arithmetic operations are related to each other.
* Develop and apply number theory concepts (e.g., primes, factors, and multiples) in real-world and mathematical problem situations.

Standard 7: Computation and Estimation
* Compute with whole numbers, fractions, decimals, integers, and rational numbers.
* Develop, analyze, and explain procedures for computation and techniques for estimation.
* Select and use an appropriate method for computing from among mental arithmetic, paper-and-pencil, calculator, and computer methods.
* Use computation, estimation, and proportions to solve problems.
* Use estimation to check the reasonableness of results.

Multiplication Bingo

In this multiplication game, students team up to practice facts to 9 and find the products on their bingo cards.

⟳→ Directions

1. Ask students to form two teams. Have them sit on opposite sides of the room.

2. Students to make game boards by drawing two vertical and two horizontal lines to divide their papers into nine boxes. Then they write one of the following numbers in any order in each box: 0, 2, 3, 4, 5, 6, 7, 8, 9, 10, 12, 14, 15, 16, 18, 20, 21, 24, 25, 27, 28, 30, 32, 35, 36, 40, 42, 45, 48, 49, 54, 56, 63, 64, 72, 81.

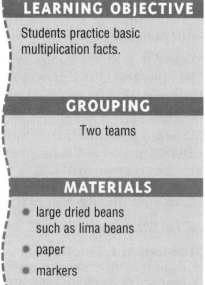

3. Say a multiplication fact to 9, such as 4 × 8 = 32, but do not give the product. Students solve the fact. If the product is on their game board, they place a bean in that box. The first student to place beans in three boxes horizontally, vertically, or diagonally wins. Allow students to play several games.

☆ Taking It Farther

Vary this game by using facts to 12.

✓ Assessing Skills

If students have difficulty solving facts, allow them to use counters to model the multiplication.

LEARNING OBJECTIVE

Students practice basic multiplication facts.

GROUPING

Two teams

MATERIALS

✳ large dried beans such as lima beans
✳ paper
✳ markers
✳ counters (optional)

Go to Bat for Facts!

Students make a hit by practicing division facts and devising strategies to win a game!

⟲→ Directions

1. Duplicate the two reproducibles for each pair of students.

2. Write the following division sentence on the board: 15 ÷ 3 = 5. Ask volunteers to identify the dividend, divisor, and quotient.

3. Explain to students that they will play a game to practice division facts.

4. Players take turns. A player chooses a number on a ball as a dividend and a number on a bat as a divisor. He or she crosses out the numbers and writes the division sentence on his or her scorecard. That player earns the same number of runs as the quotient.

5. If remaining numbers do not form basic division facts, students may write their own division facts using the dividends or divisors. For instance, if the numbers 40 and 7 were left, students could write 40 ÷ 5 = 8 or 63 ÷ 7 = 9.

6. After all the numbers are used, players calculate their total number of runs. Have them use calculators as necessary.

★ Taking It Farther

Encourage students to play again. Ask them to consider strategies they can use to achieve higher scores.

✓ Assessing Skills

Observe whether students understand the terms *dividend* and *divisor* and use the terms correctly in their division sentences.

LEARNING OBJECTIVE

Students practice basic division facts.

GROUPING

Pairs

MATERIALS

* *Go to Bat for Facts!* reproducibles (pp. 10–11)
* calculators (optional)

Go to Bat for Facts!

Choose a number on a ball as a dividend. Then choose a number on a bat as a divisor. Cross out these numbers. Write and solve the division sentence on your scorecards. You earn the same number of runs as the quotient. Take turns and then tally your final scores!

DIVIDENDS

24 47 72 16 45 40

54 36

32 12

30 36 24 25 18 21 48

30

SCORECARD: Player 1

DIVISION SENTENCE	NUMBER OF RUNS SCORED

Go to Bat for Facts!

DIVISORS

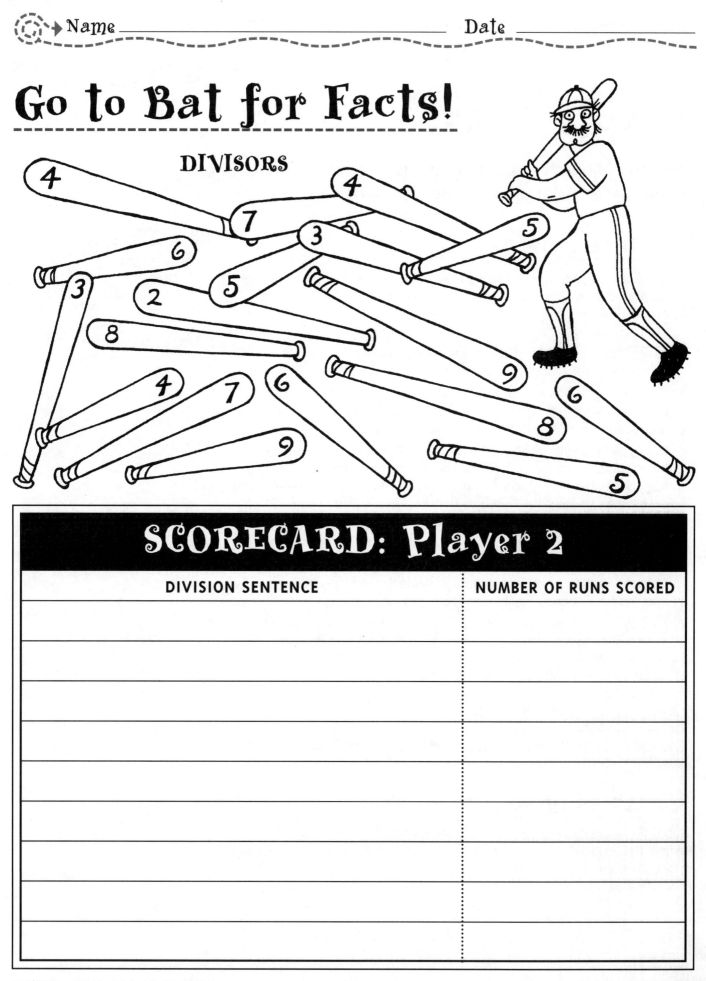

SCORECARD: Player 2	
DIVISION SENTENCE	NUMBER OF RUNS SCORED

Triple Toss

Students play a fast-paced game to review the properties of multiplication and division.

⟳➤ Directions

1. Use the masking tape to cover the dots on one face of two different number cubes. Give three cubes to each player, making sure each has a cube with one face covered. Explain that the "blank" face represents zero.

2. Review the Identity Property and the Property of Zero, and the Associative Property of Multiplication.

3. Explain to students that first they will play a multiplication game with six rounds. They will write multiplication sentences whose products result in the highest possible answers.

4. Designate students as Player 1 and Player 2. Both players put the three cubes in a cup. Each player shakes and rolls the cubes. Player 1 quickly decides if they should "switch" and exchange the cubes with each other, or keep the numbers that each has rolled.

5. Each player writes a multiplication sentence using three factors. They should use all three numbers shown, and then multiply.

6. The game continues for six rounds, with students taking turns calling "switch" or "keep." After six rounds, students add the products to find the total scores. The higher score wins the game.

7. To play the division game, players take turns tossing all six number cubes. They use two of the cubes as the dividend and divisor in a division sentence. The object of this game is to get the highest score possible. Play continues for six rounds.

✪ Taking It Farther

Have students play the multiplication and division games. This time, the lower final score wins the game.

✔ Assessing Skills

Watch for students who need more work multiplying three factors. Review the Associative Property by asking students to group the factors in two different ways. As necessary, use counters to illustrate the Identity Property and the Property of Zero.

LEARNING OBJECTIVE

Students use the properties of multiplication and division.

GROUPING

Pairs

MATERIALS

For each pair:
✷ 6 number cubes
✷ masking tape
✷ 2 small cups
✷ counters (optional)

Find the Factors

Students will have fun testing their concentration as they look for factors of certain products.

◉→ Directions

1. Write each of the following numbers in red on index cards: 1, 3, 3, 4, 4, 5, 5, 6, 6, 7, 7, 8, 8, 12, and 16. Write each of the following numbers in blue on index cards: 24, 42, 45, 48, and 56.

2. Ask students to form two teams. Have them sit on opposite sides of the room.

3. Shuffle the red cards and place them facedown to form a 5-by-3 grid. These are the factor cards. Shuffle the blue cards and place them to the right of the other cards to form a final 5-by-4 grid. These are the product cards. Leave some space between the sets of cards to distinguish them.

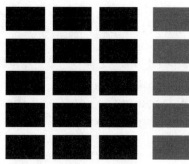

4. The first player turns over two factor cards and one product card. The student says whether the factors give that product. If the factors do give that product, and the student answers correctly, then the team takes the two factor cards and turns the product card facedown again. The next player on the same team turns over two factor cards and one product card in a similar manner. If the factors do not give the product, he or she turns all the cards over again. Play goes to the other team.

5. Continue until all factor cards are gone. The team with more factor cards wins.

★ Taking It Farther

Depending on the ability of the class, you can vary this game by using more difficult multiplication facts.

✓ Assessing Skills

Are students having difficulty matching factors with the products? You may want to have them write the factors of each product on the board.

LEARNING OBJECTIVE

Students recognize factors for specific numbers.

GROUPING

Two teams

MATERIALS

* large index cards
* red and blue markers

Factor Facts

The fact is that this multiplication game reinforces the concept of factors and related facts.

⟳→ Directions

1. For each group, make a set of cards listing numbers that students are able to factor. Numbers may include 10, 12, 16, 18, 20, 24, 30, 36, and 45.

2. Review the concept of factor pairs—numbers that, when multiplied, give the same product. Write the number 12 on the board and ask students to list all of the factor pairs. [1×12, 12×1, 2×6, 6×2, 3×4, 4×3] Then choose one of the factor pairs such as 2 and 6, and ask a volunteer to write a multiplication and division fact family for the numbers. [$2 \times 6 = 12$, $6 \times 2 = 12$, $12 \div 2 = 6$, $12 \div 6 = 2$]

3. Explain to students that they will be playing a game involving factors. They begin the game by stacking the cards and then turning over the top card. Each player in the group writes a factor pair for that number and the fact family. Then players show their pairs. An example of factor pairs for the number 12 is given below.

Player A	3, 4
Player B	6, 2
Player C	3, 4

4. Players score 1 point for each correct factor pair and 2 points for a correct fact family. (You may want to provide calculators for students to check their answers.) Players score 3 points for factor pairs that are not listed by any other player. In the above example, Players A and C score 1 point, while Player B scores 3 points for factor pairs. If all players wrote a complete and correct fact family, each would score 2 additional points.

5. Students continue playing until all of the cards are used. The highest score wins the game.

☆ Taking It Farther

Provide students with one factor and the product, such as 3 and 12. Challenge them to supply the missing factor.

✓ Assessing Skills

If students have difficulty finding factors, provide them with counters. They may use the counters to model the given number and make equal groups to show the factors. Make sure students understand that the number of counters in a group and the number of groups are factors.

LEARNING OBJECTIVE

Students find factors for a specific number and write the fact family.

GROUPING

Cooperative groups of 3

MATERIALS

✻ index cards
✻ marker
✻ paper and pencil
✻ calculators (optional)
✻ counters (optional)

That's a Fact!

Students clear the decks and search for fact families.

⟿ Directions

1. Distribute a deck of cards to each group. Have students remove all face cards and aces from their decks, leaving the cards 2 through 10 in all four suits, and then shuffle the cards well. They place the cards facedown in the center of the group.

2. Explain that they will be using the number cards to play a fact family game. Players take turns turning over two cards and saying a related fact for the numbers. For instance, if a player draws a 2 and an 8, the related fact might be $2 \times 4 = 8$, $2 \times 8 = 16$, or $16 \div 2 = 8$.

3. The other group members write another related fact on a sheet of paper. If there are duplicate related facts, the player who turned over the cards must complete the fact family. That player receives 1 point for each related fact he or she contributes.

4. Play continues for five rounds. Groups may reshuffle the cards as necessary.

★ Taking It Farther

Let students play the game verbally. Each takes a turn saying a related fact. Encourage them to say the facts as quickly as they can.

✓ Assessing Skills

Do students understand the relationship among the three numbers in the fact family?

LEARNING OBJECTIVE

Students write fact families for groups of three numbers.

GROUPING

Cooperative groups of 4

MATERIALS

✳ 1 deck of cards for each group

✳ paper and pencil

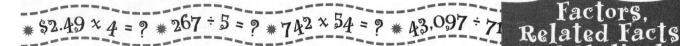

Don't Four-Get Multiples!

To solve this puzzle, students reinforce the concept of multiples and geometry.

➤ Directions

1. Make a copy of the reproducible for each student. Review the concept of multiples: The multiples of a number are the products of that number and other numbers. Ask students to list multiples for the number 3. [3, 6, 9, 12, 15, 18, 21, 24, 27, 30, and so on]

2. Then have students look carefully at the reproducible to find triangles within the large triangle containing dots that add up to a multiple of 4. Point out that the triangles may be of any size.

3. Invite pairs of students to compare their results.

⭐ Taking It Farther

Encourage students to make similar puzzles with multiples of 5. They may exchange puzzles with classmates.

☑ Assessing Skills

* If students have difficulty finding multiples of 4, display a chart before they begin.

* If students are unable to visualize the different triangles, you may want to have them begin by outlining triangles that they see with a marker.

LEARNING OBJECTIVE

Students review multiples, focusing on the number 4.

GROUPING

Individual

MATERIALS

* *Don't Four-Get Multiples!* reproducible (p. 17)
* markers

ANSWERS

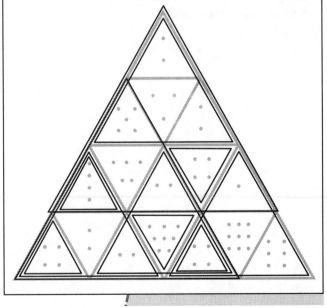

Don't Four-Get Multiples!

There are multiple ways to solve this multiple puzzle!
But you'll need to sharpen your eyes and look carefully.

Think about how to find multiples for the number 4. Then find triangles in the large triangle that contain dots adding up to a multiple of 4. The triangles may be of any size—even upside down!

Use a marker to show your triangles. Then compare your results with those of a friend.

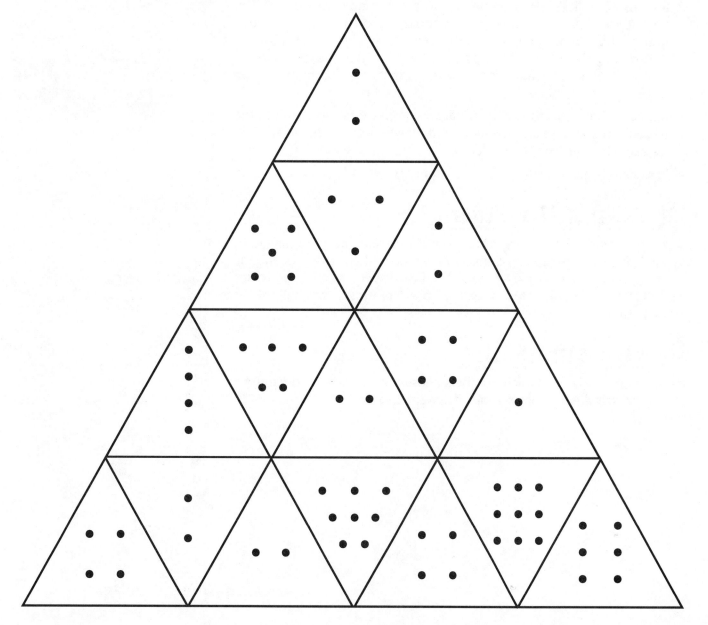

Multiplication Buzz

Students will enjoy this multiplication counting game as they recognize multiples of numbers.

⊚→ Directions

1. Have students sit in a large circle. Ask them to choose a number to be "It." Explain that they cannot name "It" or any multiple of "It." For example, if 3 is "It," then they must substitute the word "buzz" for all 3s, all multiples of 3, or whenever a 3 appears.

2. Suppose 3 is "It." Ask someone to start the game by saying "1." The next student says "2," the next says "buzz," and so on around the group. The counting continues as follows: 4, 5, buzz (2 × 3 = 6), 7, 8, buzz (3 × 3 = 9), 10, 11, buzz (4 × 3 = 12), buzz (3 appears in 13), and so on. You may wish to have students say the multiplication fact after saying "buzz."

3. A student who fails to say "buzz" at the proper time, or says it at the wrong time, or says an incorrect number, is out of the game and leaves the circle. The last student remaining is the winner.

4. Play several games, using different numbers for "It."

✪ Taking It Farther

As you go around the circle in order, some students may count ahead to prepare for their turn. To avoid this, you can have them toss a beanbag or ball to select the next player. Thus, the first player says "1" and tosses the ball to another student who says "2" and then tosses the ball to another student.

✔ Assessing Skills

Ask students to explain which numbers chosen as "It" will produce the most buzzes and why. For example, if 1 were chosen as "It," what would happen?

LEARNING OBJECTIVE

Students recognize multiples of numbers.

GROUPING

Whole class

MATERIALS

✳ beanbag or ball (optional)

Greetings from Space

Students create numbers to multiply and complete a spacey story.

⟳→ Directions

1. Duplicate and distribute a reproducible to each student. Make sure students understand that they write one digit in each space below the blank, creating 1-, 2-, or 3-digit numbers.

2. After students create numbers, they exchange stories and multiply. Discuss the patterns that occur when multiplying by multiples of 10.

3. Encourage students to read the stories aloud to each other. Talk about the similarities and differences in the completed stories.

✰ Taking It Farther

Ask partners to work together to write similar types of letters about fantasy vacations. They then exchange letters and solve.

✓ Assessing Skills

✷ Determine whether students are able to identify patterns when multiplying.

✷ Observe whether they write the correct number of zeros in the products.

LEARNING OBJECTIVE
Students explore patterns when multiplying by multiples of 10.

GROUPING
Pairs

MATERIALS
✷ *Greetings from Space* reproducible (p. 20)

Name _____ Date _____

Greetings from Space

Harry had a strange trip this summer.

Help him write a postcard to his friend. Write a 1-digit, 2-digit, or 3-digit number in the space(s) below each blank. Then give to a friend to solve.

Hi Mel!
Greetings from deep space! Having a great time! Wish you were here!

To: Mel
2315 Ash
Bridge...

Hi, Mel! This space trip has been totally strange. The first day, we traveled _____ miles. We landed on the planet Mojomo,
(_ _ × 20)

and about _____ aliens were there. We must have looked
(100 × _ _)

hungry, because they brought us about _____ sandwiches.
(30 × _ _ _ _)

We didn't want to be rude, so we ate as many as we could.

Then we were tired, so we slept for _____ hours under
(10 × _ _)

a beautiful tree. But when we woke up, there were about

_____ wild animals circling around us! We didn't want to be
(100 × _)

their lunch, so luckily we had some sandwiches left to give them.

We ran to our spaceship and traveled back to Earth. As they

say, there's no place like home!

20

Multiplication and Division Scholastic Professional Books

✳ $2.49 × 4 = ? ✳ 267 ÷ 5 = ? ✳ 742 × 54 = ? ✳ 43,097 ÷ 7...

Exploring Patterns (Multiplication)

Dot to Dot

This game utilizes patterns to sharpen students' mental math skills.

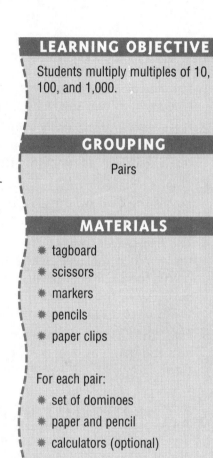

⟲➔ Directions

1. Make a spinner out of tagboard like the one shown for each pair. Divide each spinner into three equal sections. labeled 0, 00, and 000. Distribute the spinners, dominoes, and calculators (optional) to pairs of students.

2. Explain to students that they will play a multiplication game. The object is to score the greatest number of points.

3. Players make an array of dominoes facedown. The first player turns over a domino and writes the two numbers represented by dots on its face. The player chooses one number to be the first factor. To form the second factor, she or he spins the spinner and writes that number of zeros after the other number.

4. The player mentally multiplies the factors and says the product. If both players agree that the product is correct, the player records the product. The product is the player's score for that round.

5. Students take turns. After five rounds, they add to find their total scores. You may want to have students use calculators to check the scores for each round as well as the total scores.

☆ Taking It Farther

Talk about strategies students developed to gain higher scores. Then let them play again. This time, the lower score wins the game. Compare the strategies students used to play each game.

✓ Assessing Skills

Assess students' ability to use mental math to multiply. If they seem unsure at any stage, write the following on the board:

$$4 \times 3$$
$$4 \times 30$$
$$4 \times 300$$

Allow them to use models to understand the relationships.

LEARNING OBJECTIVE

Students multiply multiples of 10, 100, and 1,000.

GROUPING

Pairs

MATERIALS

* tagboard
* scissors
* markers
* pencils
* paper clips

For each pair:
* set of dominoes
* paper and pencil
* calculators (optional)

Star Cards

Are students starstruck? They'll find out when they play this fast-paced division game.

⟳→ Directions

1. Shuffle each group of cards. Place the divisor and dividend cards facedown in two stacks on a desk or table.

2. Have the class count off 1 and 2 to form two teams. Designate one student as the scorekeeper and another student as the checker. Each team stands in a line before the stacks of cards.

3. The first player on each team takes the top card from one stack and places the card on the chalkboard ledge. The two players divide mentally and write the quotient on the board.

4. The first player to write the correct quotient scores 1 point for the team. When a player picks a card with a star, he or she scores 2 points for a correct answer for that round. The checker verifies the accuracy of the answers with a calculator. The scorekeeper uses tallies to record the scores on the board.

5. Place the cards in a discard pile, reshuffling and incorporating them into the game as necessary. Play continues until everyone has had a turn.

☆ Taking It Farther

Have students make additional cards and add them to the stacks. Groups of three students can play the game. The roles of the scorekeeper and checker are combined; students rotate the roles with each turn.

✔ Assessing Skills

Note students who are experiencing difficulty with multiples. If necessary, pause during the game and provide some practice equations, relating the number of zeros in the factors to the number of zeros in the product.

LEARNING OBJECTIVE

Students divide multiples of 10, 100, and 1,000.

GROUPING

Two teams

MATERIALS

＊ large index cards
＊ markers
＊ calculators
＊ chalk

ADVANCE PREPARATION

Write the following divisors on index cards, then stack them: 3, 3, 4, 4, 6, and 6. Make another set of cards with the following dividends: 12; 24; 120; 240; 1,200; 2,400; 12,000; and 24,000. Draw a star in a corner of one of the divisor cards and on two of the dividend cards.

Can You Dig It?

If someone asked you to "dig it," would you get a shovel? For this game, students use front-end estimation and discover some '60s phrases.

◉→ Directions

1. Discuss how the same words can mean different things to different people and how words change meanings at different times. Ask students to share some slang expressions that they use.

2. Review how to use front-end estimation using the example below.

Multiply. 37 × 281	Use front-end estimation. 30 × 200	The estimate is 6,000.

3. Partners place the cards facedown in a 4-by-4 grid and play a Concentration-type memory game. The first player turns over two cards and uses front-end estimation to multiply any multiplication problem. The object is to match the multiplication problem ('60s slang) with the estimated product (its definition). Encourage students to try to remember the location of the estimated products. [Pad, Home (46 × 389; estimate, 12,000); Hairy, Scary (56 × 89; estimate, 4,000); Do a number, Persuade (426 × 7; estimate, 2,800); Scene, Where the fun is (295 × 471; estimate, 80,000); Hang me up, Keep me waiting (97 × 51; estimate, 4,500); Get down, Have fun (628 × 43; estimate, 24,000); Trash, Destroy (578 × 382; estimate, 150,000); Flower Power, Peace and love (87 × 7; estimate, 560)]

4. Players keep matches, and the player with more cards wins. Groovy!

✪ Taking It Farther

Let students create their own '90s version of the game, using contemporary slang for the multiplication problems and definitions for the estimated products.

✓ Assessing Skills

Note whether students have trouble relating place value to front-end estimation. Have them use place-value charts to estimate.

LEARNING OBJECTIVE

Students estimate products using front-end estimation.

GROUPING

Pairs

MATERIALS

✳ index cards
✳ markers

ADVANCE PREPARATION

Make a set of 16 index cards with the following written on them:

Pad 46 x 389	Hairy 56 x 89
Do a number 426 x 7	Scene 295 x 471
Hang me up 97 x 51	Get down 628 x 43
Trash 578 x 382	Flower Power 87 x 7
Have fun 24,000	Destroy 150,000
Keep me waiting 4,500	Peace and love 560
Persuade 2,800	Home 12,000
Scary 4,000	Where the fun is 80,000

Riddle Round-Up

Estimating can be a laughing matter when students round and divide.

➔ Directions

1. Write 267 ÷ 5 on the board. Review and discuss how to estimate the quotient by rounding and using compatible numbers.

ROUNDING	COMPATIBLE NUMBERS
Round dividend to the greatest place and divide. 267 ÷ 5 ↓ 300 ÷ 5 = 60	Use basic division facts. 267 ÷ 5 ↓ 250 ÷ 5 = 50

2. Give partners a copy of the reproducible and have them cut out and glue a riddle and division problem to the front of an index card. Then they cut out the quotient and answer. Students should fold the quotient/answer so that the quotient appears on one side of the index card and the answer to the riddle is on the back. The riddle and division problem go on one card. The answer and quotient appear on a separate card.

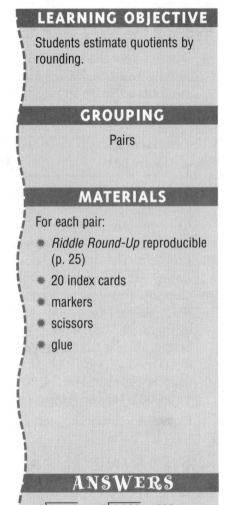

5,000

3. One student chooses a riddle card and estimates the quotient by using compatible numbers. The partner finds the matching card with the estimated quotient and reads the answer to the riddle. Students use the answer to the riddle to make sure the estimate is sensible.

4. Students take turns and continue until all the riddles have been solved.

☆ Taking It Farther

Invite students to choose some of the problems and estimate the quotients by rounding. Have them compare the estimates derived by both methods. Talk about situations where each type of estimation might be appropriate.

✓ Assessing Skills

✻ Do students use the closest basic division fact?

✻ Are they putting the correct number of zeros in the quotients?

LEARNING OBJECTIVE

Students estimate quotients by rounding.

GROUPING

Pairs

MATERIALS

For each pair:

✻ *Riddle Round-Up* reproducible (p. 25)

✻ 20 index cards

✻ markers

✻ scissors

✻ glue

ANSWERS

$8\overline{)3{,}719} \rightarrow 8\overline{)4{,}000} = 500$

$57 ÷ 3 \rightarrow 60 ÷ 3 = 20$

$711 ÷ 9 \rightarrow 720 ÷ 9 = 80$

$7\overline{)462} \rightarrow 7\overline{)490} = 70$

$5{,}680 ÷ 6 \rightarrow 5{,}400 ÷ 6 = 900$

$5\overline{)16{,}102} \rightarrow 5\overline{)15{,}000} = 3{,}000$

$4\overline{)2{,}620} \rightarrow 4\overline{)2{,}400} = 600$

$42{,}888 ÷ 9 \rightarrow 45{,}000 ÷ 9 = 5{,}000$

$8\overline{)741} \rightarrow 8\overline{)720} = 90$

$3{,}707 ÷ 9 \rightarrow 3{,}600 ÷ 9 = 400$

Riddle Round-Up

Estimating can be a laughing matter!

Cut out and glue a riddle and division problem to the front of an index card. Cut out, fold, and glue a quotient and answer to the front and back of an index card.

Now you're ready to play! Estimate by *using compatible numbers*, find the matching estimation card, and see if you can answer the riddles.

Riddles and Division Problems

Why did the man bring a rope to the baseball game?	$8\overline{)3,719}$
What does a shark eat with peanut butter?	$57 \div 3$
What has four legs and flies?	$711 \div 9$
What did one eye say to the other?	$7\overline{)462}$
What do you call the time of prehistoric pigs?	$5,680 \div 6$
Why can't a bicycle stand by itself?	$5\overline{)16,102}$
What kind of jam cannot be eaten?	$4\overline{)2,620}$
What did one math book say to the other?	$42,888 \div 9$
When is a black dog not a black dog?	$8\overline{)741}$
Why did the boy put an alarm clock in his shoe?	$3,707 \div 9$

Quotients and Answers

20	5,000
jellyfish	Boy, I've got problems!
900	400
Jurassic Pork	so his foot wouldn't fall asleep
3,000	500
because it's two tired	to tie up the game
80	90
a horse	when it's a greyhound
70	600
There's something between us that smells.	a traffic jam

Productive Draws

Students sharpen critical-thinking skills and decide which factors will give the highest product.

⟶ Directions

1. Write the multiplication sentence 35 x 9 on the board. Review the regrouping process as follows:

Multiply the ones. Regroup if necessary.	Multiply the tens. Add any new tens.
4	4
35	35
× 9 (9 × 5 = 45)	× 9 (9 × 3 = 27)
5	315 (27 + 4 = 31)

2. Ask students to form Team A and Team B. Have teams sit on opposite sides of a table.

3. Explain that the object of the game they will play is to use factors to make the highest product possible.

4. Shuffle the cards and place them facedown in a stack on the table. The first player on each team draws three cards. The first player on Team A places two cards faceup on the table to form a 2-digit factor of his or her choice. Then the first player on Team B places one card below the two to form a multiplication problem.

5. Students multiply to find the product. Team A records the product as its first score. Players place their cards in a discard pile.

6. Repeat the process, with the second player on Team B going first to form a 2-digit factor and the second player on Team A going second to select a 1-digit factor. Team B records the product as its score. Play continues until all students have drawn cards. (You may need to reshuffle the discard pile before the game ends.) Teams add their scores. They may use calculators if you wish. The team with the higher score wins.

☆ Taking It Farther

For an additional challenge, the second player of each round can place two cards below the first factor to form a 2-digit factor. You can also have students draw five cards and form 3- and 2-digit factors.

✓ Assessing Skills

✴ Observe what kinds of strategies students are developing.

✴ Ask them to explain how to obtain the highest scores.

LEARNING OBJECTIVE

Students use logical reasoning and practice multiplying by a 1-digit number.

GROUPING

Two teams

MATERIALS

✴ deck of cards

✴ paper and pencil

✴ calculator (optional)

ADVANCE PREPARATION

Remove all tens and face cards from a deck of cards. The number cards 2 through 9 and the aces should remain.

26

Time to Grow Up!

To use this bulletin board, students match animals with their gestation or incubation periods. Then they create multiplication problems.

➔ Directions

1. Make the bulletin board as shown below. On the back of the cards showing the gestation time, draw a picture of the animal or write its name. The times listed are averages.

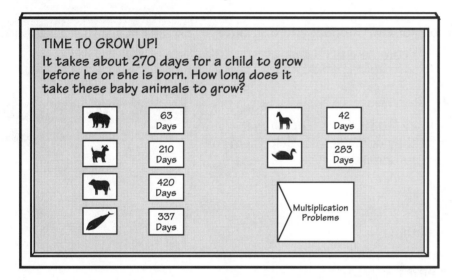

2. You may want to explain that animals can grow inside their mothers, which is called gestation, or they can grow in eggs outside their mothers, which is called incubation.

3. To begin, students guess and match the animals and their growing times. They can check their guesses by looking at the back of the cards.

4. After students match, they make up multiplication problems, such as the following: *If a mother horse is pregnant 2 times in her life, how many days will she be pregnant?* [674 days] Challenge students to solve each other's problems.

⭐ Taking It Farther

Invite students to research the gestation or incubation times for other animals. Let them include these animals on the bulletin board.

✓ Assessing Skills

Make sure students are using the correct numbers to solve the problems.

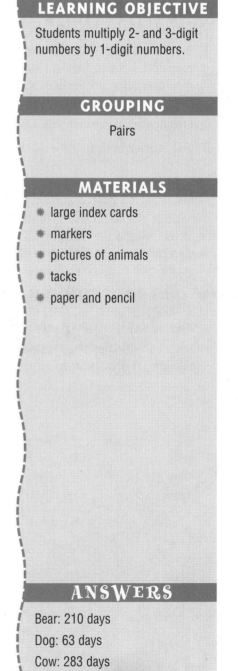

LEARNING OBJECTIVE

Students multiply 2- and 3-digit numbers by 1-digit numbers.

GROUPING

Pairs

MATERIALS

✳ large index cards
✳ markers
✳ pictures of animals
✳ tacks
✳ paper and pencil

ANSWERS

Bear: 210 days

Dog: 63 days

Cow: 283 days

Whale: 420 days

Horse: 337 days

Swan: 42 days

Believe It or Not!

Students find out interesting facts by multiplying and completing some strange sentences.

→ Directions

1. Review multiplying with regrouping by writing the following example on the board: 3 × 39. Have students write the multiplication in vertical form and solve. Work through the regrouping process using models if necessary.

2. Duplicate a copy of the reproducible for each student and distribute.

3. Tell students that they will multiply to complete each fact.

☆ Taking It Farther

Talk about the facts that are most surprising to students. Encourage them to make a list of unusual facts that they come across. Help them to create equations to fill in the blanks as appropriate.

✓ Assessing Skills

* Note whether students use mental math.
* Observe whether they make common errors such as forgetting to regroup, regrouping incorrectly, or errors with basic facts.

LEARNING OBJECTIVE

Students multiply 2-digit and 3-digit numbers by 1-digit numbers.

GROUPING

Individual

MATERIALS

* *Believe It or Not!* reproducible (p. 29)

ANSWERS

1. 10 days
2. 100 miles per hour
3. 70 miles an hour
4. 38 pounds
5. 68 miles per hour
6. 360 calories
7. 150 feet
8. 1,100 million people
9. 639 muscles; 40 percent
10. 280 miles an hour

Believe It or Not!

Chances are you won't know the answers to these wacky facts! Multiply and then write the product in the blank. What you find out may surprise you!

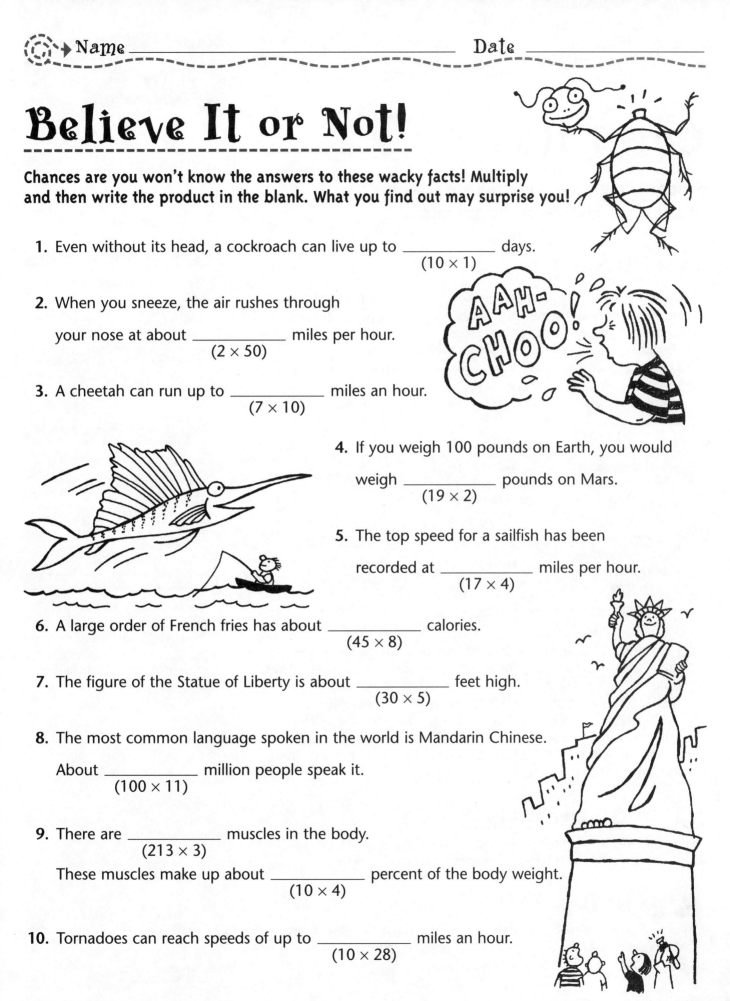

1. Even without its head, a cockroach can live up to _____ days.
(10 × 1)

2. When you sneeze, the air rushes through

    your nose at about _____ miles per hour.
    (2 × 50)

3. A cheetah can run up to _____ miles an hour.
(7 × 10)

4. If you weigh 100 pounds on Earth, you would

    weigh _____ pounds on Mars.
    (19 × 2)

5. The top speed for a sailfish has been

    recorded at _____ miles per hour.
    (17 × 4)

6. A large order of French fries has about _____ calories.
(45 × 8)

7. The figure of the Statue of Liberty is about _____ feet high.
(30 × 5)

8. The most common language spoken in the world is Mandarin Chinese.

    About _____ million people speak it.
    (100 × 11)

9. There are _____ muscles in the body.
(213 × 3)

    These muscles make up about _____ percent of the body weight.
    (10 × 4)

10. Tornadoes can reach speeds of up to _____ miles an hour.
(10 × 28)

Car Talk

Students multiply and create a timeline to show the history of the automobile.

⟳→ Directions

1. Write the following facts on one side of the index cards and the multiplication problems on the back. Have students multiply and write the products on the front beside the event. The products will show the years in which the events took place.

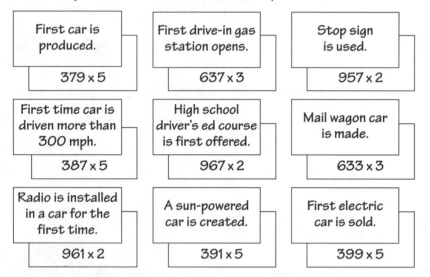

First car is produced.	First drive-in gas station opens.	Stop sign is used.
379 × 5	637 × 3	957 × 2

First time car is driven more than 300 mph.	High school driver's ed course is first offered.	Mail wagon car is made.
387 × 5	967 × 2	633 × 3

Radio is installed in a car for the first time.	A sun-powered car is created.	First electric car is sold.
961 × 2	391 × 5	399 × 5

2. Help students tape the cards on the appropriate place on the timeline as necessary. Encourage them to draw illustrations for the timeline. [1895—First car is produced. 1899—Mail wagon car is made. 1911—First drive-in gas station opens. 1914—Stop sign is used. 1922—Radio is installed in a car for the first time. 1934—High school driver's ed course is first offered. 1935—First time a car is driven more than 300 mph. 1955—A sun-powered car is created. 1995—First electric car is sold.]

☆ Taking It Farther

Invite students to research other facts about the history of the automobile. Have them include these facts on the timeline.

✓ Assessing Skills

Observe if students place the dates correctly on the timeline. Point out how a timeline is like a number line. If necessary, have students practice placing 2-digit numbers on number lines.

LEARNING OBJECTIVE

Students multiply 3-digit numbers by 1-digit numbers.

GROUPING

Small cooperative groups

MATERIALS

* chart paper
* index cards
* markers
* tape

ADVANCE PREPARATION

Use chart paper and markers to make a timeline showing ten-year increments from 1895 through 1995. Leave ample space so that students can tape index cards on the timeline.

Weigh up in Space

Students multiply to learn about size and weight of planets as they help to make a bulletin board.

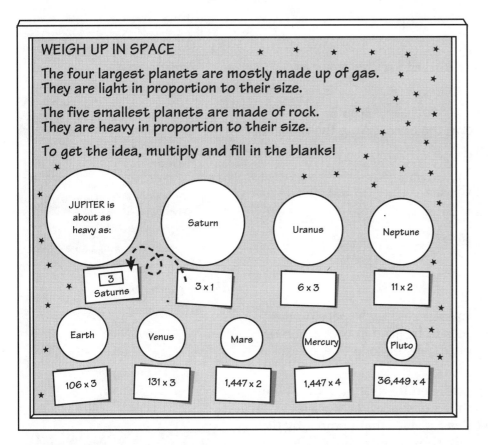

WEIGH UP IN SPACE

The four largest planets are mostly made up of gas.
They are light in proportion to their size.

The five smallest planets are made of rock.
They are heavy in proportion to their size.

To get the idea, multiply and fill in the blanks!

JUPITER is about as heavy as:

Saturn	Uranus	Neptune

3 Saturns | 3 x 1 | 6 x 3 | 11 x 2

Earth | Venus | Mars | Mercury | Pluto

106 x 3 | 131 x 3 | 1,447 x 2 | 1,447 x 4 | 36,449 x 4

⟳→ Directions

Students multiply and write the products in the rectangles on the back of the cards. Then they reattach the cards to the appropriate planets. You may want to have them check their work by using a calculator. [Jupiter is about as heavy as 3 Saturns; 18 Uranuses; 22 Neptunes; 318 Earths; 393 Venuses; 2,894 Marses; 5,788 Mercuries; 145,796 Plutos.]

☆ Taking It Farther

Invite students to find out the diameter in miles of each planet. Have them order the numbers and make up problems for others to solve.

✓ Assessing Skills

Ask students to describe the regrouping process when multiplying 4- and 5-digit numbers.

LEARNING OBJECTIVE

Students multiply by 1-digit numbers.

GROUPING

Small cooperative groups

MATERIALS

* bulletin board
* markers
* index cards
* art paper
* tacks
* calculator (optional)

ADVANCE PREPARATION:

Prepare the bulletin board as shown. Cut out proportionate circles to show the planets and label them accordingly. Write the matching multiplication problem on an index card. On the other side of the card draw a rectangle followed by the planet's name in plural (for example, Earths, Neptunes). Place the card under the appropriate planet.

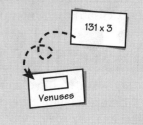

What Do You Know?

Students think of numbers they already know as they play a division game.

⟶ Directions

1. Write the example shown on the board and review the steps involved in the division.

Step 1
Determine where to place the first digit in the quotient. Divide.

4 > 1
4 < 15   Divide tens.

$$4\overline{)159}\\ \underline{-12}\\ 3$$
(quotient: 3)

Step 2
Bring down the ones. Divide.

$$4\overline{)159}\ 39\\ \underline{-12}\\ 39\\ \underline{-36}\\ 3$$

Step 3
Write the remainder.

$$4\overline{)159}\ \ 39\ R3\\ \underline{-12}\\ 39\\ \underline{-36}\\ 3$$

2. Give a copy of the reproducible to each pair. Each space on the game board has a square that needs to be filled in with a number. The number can be part of a book, movie, or song title, or can complete a phrase.

3. Players use counters for game pieces, and the first player rolls the number cube and moves that number of spaces. That player writes the correct number in the square and divides that number by the number represented on the number cube. For example, if Player 1 rolls 4, he or she divides 13 by 4.

4. The partner checks the division. If the division is correct, the player stays in the space. If it is incorrect, he or she moves back to START.

5. Players must roll the exact number to reach FINISH.

☆ Taking It Farther

Use correction tape to make some blank spaces on the game board. Have students create their own phrases for the game and play again.

✓ Assessing Skills

You may want to have some students fill in the blanks on the game board before playing the game. Then these students can focus on the division as they play.

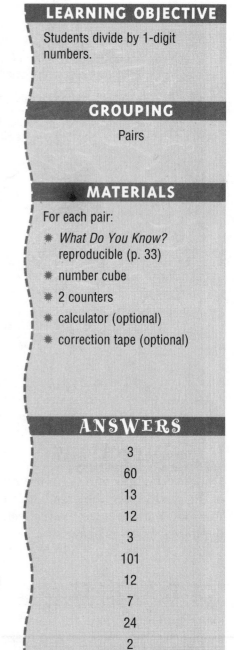

LEARNING OBJECTIVE

Students divide by 1-digit numbers.

GROUPING

Pairs

MATERIALS

For each pair:
✻ *What Do You Know?* reproducible (p. 33)
✻ number cube
✻ 2 counters
✻ calculator (optional)
✻ correction tape (optional)

ANSWERS

3
60
13
12
3
101
12
7
24
2
12
20,000
4
8

What Do You Know?

You'll need to think of numbers you already know to play this division game!

Roll the number cube and then move that many spaces on the game board. Fill in the square with the correct number. Then make and solve a division sentence with the number you rolled and the number in the square.

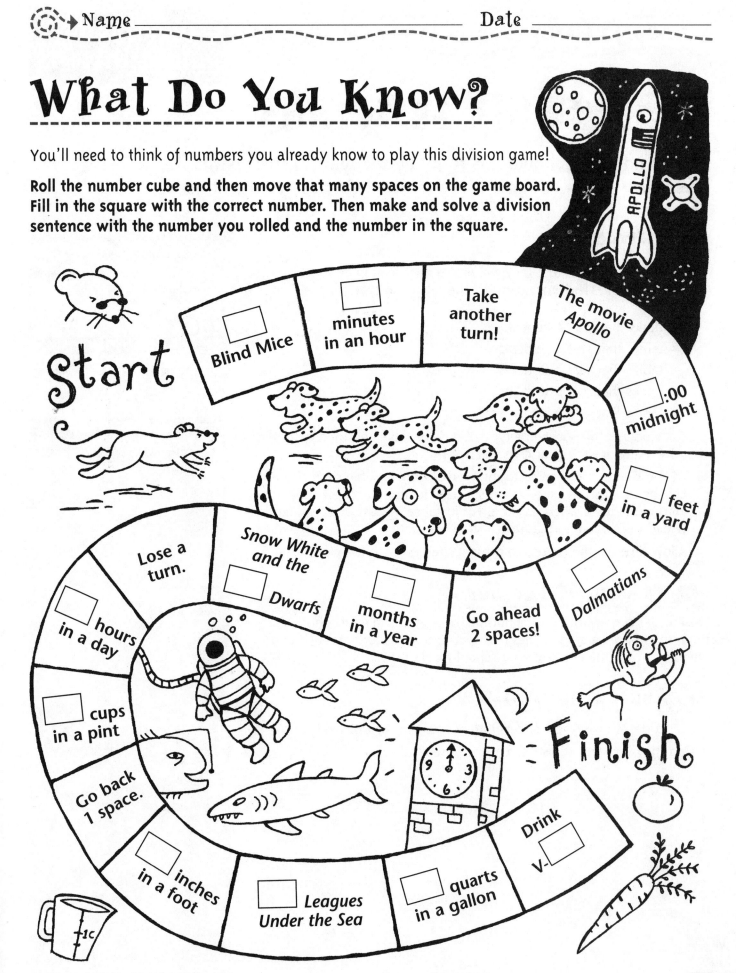

Start

☐ Blind Mice

☐ minutes in an hour

Take another turn!

The movie Apollo ☐

☐ :00 midnight

☐ feet in a yard

☐ Dalmatians

Go ahead 2 spaces!

☐ months in a year

Snow White and the ☐ Dwarfs

Lose a turn.

☐ hours in a day

☐ cups in a pint

Go back 1 space.

☐ inches in a foot

☐ Leagues Under the Sea

☐ quarts in a gallon

Drink ☐ V-

Finish

Your Table Is Ready

There's no need for students to fly by the seats of their pants when they figure out how many tables their classroom will need.

⊙→ Directions

1. Explain to students that they will plan a new seating arrangement for the classroom. They may consider the following types of desks and tables:
   - ✳ desks that seat 2 students
   - ✳ tables that seat 3 students
   - ✳ tables that seat 4 students
   - ✳ tables that seat 5 students
   - ✳ tables that seat 6 students

2. Ask students to choose one type of desk or table and then determine how many of each your class will need.

3. Divide students according to the type of desk or table each chose. Let the groups compare work, especially focusing on how they dealt with remainders when dividing. For instance, a class of 32 students would need 7 tables that seat 5 students each.

4. Allow time for the groups to present their ideas to the class.

✪ Taking It Farther

Present the class with the total number of students in your school. Ask them to choose cafeteria tables that seat 4, 6, or 8 students each and determine how many tables your school would need.

✓ Assessing Skills

✳ Are students taking remainders into account in their orders?

✳ Can they verbalize the strategies they used?

LEARNING OBJECTIVE

Students divide 2-digit numbers by 1-digit numbers and work with remainders.

GROUPING

Individuals and then groups

MATERIALS

✳ paper and pencil

Career Changes

Students match a president of the United States with a former career and practice division skills.

⟳→ Directions

1. Duplicate the *Career Changes* reproducible for each student.

2. Have students complete the page individually. You may want to let pairs compare their answers and talk about facts about the presidents that surprised them.

✪ Taking It Farther

Encourage students to research the lives of other presidents. They can use the information to write their own multiplication matches for other students to solve.

✓ Assessing Skills

✻ Ascertain whether students place the first digit correctly in the quotient.

✻ Ask students why the remainder cannot be larger than the divisor.

LEARNING OBJECTIVE

Students divide 2- and 3-digit numbers by 1-digit divisors.

GROUPING

Individual

MATERIALS

✻ *Career Changes* reproducible (p. 36)

ANSWERS

1. 12 R3, Soldier
2. 48 R1, Teacher
3. 66, Actor
4. 93 R4, Newspaperman
5. 54, Tailor
6. 48 R2, Oilman
7. 93 R6, Teacher
8. 36, Peanut farmer
9. 151, Writer

Career Changes

Many Americans will change careers during their lives.
Even presidents of the United States have had different jobs!

**What careers do you think these presidents had before they moved
into the White House? Your job is to find out! First, take a guess.
Then divide, match the quotient, and see if your guess is correct.**

DIVIDE:

Jackson

1. Andrew Jackson

   $75 \div 6 =$ _____

2. Woodrow Wilson

   $193 \div 4 =$ _____

Reagan

3. Ronald Reagan

   $462 \div 7 =$ _____

4. John F. Kennedy

   $469 \div 5 =$ _____

Kennedy

5. Andrew Johnson

   $432 \div 8 =$ _____

6. George Bush

   $146 \div 3 =$ _____

7. Lyndon Johnson

   $843 \div 9 =$ _____

Jefferson

8. Jimmy Carter

   $72 \div 2 =$ _____

9. Thomas Jefferson

   $755 \div 5 =$ _____

QUOTIENT AND CAREER:

66   Actor

151   Writer

93 R6   Teacher

48 R2   Oilman

48 R1   Teacher

12 R3   Soldier

36   Peanut farmer

54   Tailor

93 R4   Newspaperman

Continue Her-story!

Students discover some amazing women in history as they practice division.

⟲→ Directions

1. Review division that results in a zero in the quotient.

2. Duplicate and distribute a copy of the reproducible to each student or pair.

3. Explain to students that they need to find the sentence in the second column to continue the story begun in the first column. After dividing the number in the first column by the number in the second column, they check the list of women at the top of the page to find the matching quotient. Do the first exercise together, reviewing the division process as necessary.

4. You may want to have students work in pairs. They can take turns matching the story sentences and doing the division.

☆ Taking It Farther

Ask each student to interview a woman to find out about something she has accomplished. Students can work in groups to prepare a similar set of exercises.

✓ Assessing Skills

Watch for students who do not find the matching quotient after they divide. Help them use logical reasoning to determine if the factors do not match, or if they made a division error.

LEARNING OBJECTIVE
Students divide 4- and 5-digit numbers by 1-digit numbers.

GROUPING
Individual or pairs

MATERIALS
✻ *Continue Her-story* reproducible (p. 38)

ANSWERS

$12,696 ÷ 3 = 4,232$
(Nellie Bly)

$2,130 ÷ 2 = 1,065$
(Sarah Edmonds)

$14,790 ÷ 5 = 2,958$
(Elizabeth Blackwell)

$18,921 ÷ 3 = 6,307$
(Ida B. Wells)

$25,914 ÷ 6 = 4,319$
(Lydia Pinkham)

Continue Her-story!

Many women have done deeds that have made history, but we don't always hear about it! Here's your chance to find out more.

Read about the deed in the first column. Then find the sentence in the second column that continues the story. Divide the number in the first column by the number in the second column and match the quotient to find the accomplished woman.

Elizabeth Blackwell 2,958
Nellie Bly 4,232
Lydia Pinkham 4,319
Ida B. Wells 6,307
Sarah Edmonds 1,065

This woman was called "the best reporter in America." (12,696)

This woman disguised herself as a man to fight in the Civil War. (2,130)

About 20,000 people came to watch her become the first woman to receive an M.D. (a doctor's degree). (14,790)

When she was 18, this woman was forced to leave a train because she wouldn't sit in the "colored only" section. (18,921)

In 1875, this woman started selling an herbal medicine and offering advice about women's health. (25,914)

Her medicine sales started earning $300,000 a year! (6)

When she couldn't find an American hospital that would hire her, she began her own hospital. (5)

She wrote a book about her war adventures that sold 175,000 copies. (2)

Once she pretended to be insane so she could write a news story about mental hospitals. (3)

She sued the railroad and continued to write about problems in the South. (3)

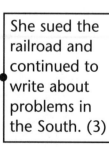

Multiplication and Division Scholastic Professional Books

You're in Jeopardy

The answers are the easy part—it's the questions that are tricky!

🕸➜ Directions

1. Review the terms *dividend*, *divisor*, and *quotient* by writing the example on the board and having students identify and label each term: 250 ÷ 7 = 35 R5.

2. Explain to students that they will play a division Jeopardy game. The answers will be given, and the players must supply the questions in the categories of "Dividends," "Divisors," and "Quotients." One student will be the game host and read the answers and questions out loud; another student will keep score; and the remaining 2 or 3 students will be contestants.

3. Distribute a copy of the reproducible to each group. On index cards, they write five questions and answers for each category. The cards should have the format shown.

QUOTIENTS $100	
Answer:	The dividend is 40.
	The divisor is 8.
Question:	What is 5?

4. Students cut apart the Jeopardy squares. On the back of the matching squares, they write only the answers from their corresponding index cards. Groups then exchange sets of index cards and Jeopardy squares.

5. The contestants take turns choosing categories and money amounts. The game host turns over the appropriate squares and reads aloud the answers. All contestants may respond by raising their hands. The host determines who raised his or her hand first and refers to the index cards for the correct questions. Contestants may use paper and pencils to solve the problems. If the contestant gives the correct question to the answer, he or she earns that amount of money. The scorekeeper records the amount. That contestant may choose another category.

6. Play continues until all the squares have been turned over and read.

★ Taking It Farther

Expand the game categories to include the multiplication terms *factors*, *products*, and *multiples*.

✔ Assessing Skills

Are students able to connect the category terms to the proper parts of the division sentence?

LEARNING OBJECTIVE

Students identify dividends, divisors, and quotients.

GROUPING

Cooperative groups of 4 or 5

MATERIALS

For each group:
* *You're in Jeopardy* reproducible (p. 40)
* 15 index cards
* markers
* scissors
* calculators
* paper and pencil

You're in Jeopardy

How about Divisors for $200? Don't forget to give your answer in the form of a question!

Cut out the game squares below. Write an answer on the back of each square.

Dividends $100	**Divisors $100**	**Quotients $100**
Dividends $200	**Divisors $200**	**Quotients $200**
Dividends $300	**Divisors $300**	**Quotients $300**
Dividends $400	**Divisors $400**	**Quotients $400**
Dividends $500	**Divisors $500**	**Quotients $500**

Multiplication and Division Scholastic Professional Books

River Rapids

Students take a "rapid" trip as they multiply
and divide to make their way through a maze.

⟳→ Directions

1. Duplicate and distribute the reproducible. Explain that the object is to follow the maze by forming correct multiplication sentences. If the path is correct, the answer to one sentence will be the first number in the next sentence.

2. Point out that if a sentence in the maze is not correct, students should try another path.

★ Taking It Farther

Pair students of similar abilities and challenge them to time each other to see how "rapidly" they can complete the maze.

✓ Assessing Skills

Determine how effectively students are able to backtrack if they make a multiplication error or go down the wrong path.

LEARNING OBJECTIVE

Students multiply by 1- and 2-digit numbers.

GROUPING

Individual

MATERIALS

✳ *River Rapids* reproducible (p. 42)

ANSWERS

7 x 3 = 21

21 x 4 = 84

84 x 1 = 84

84 x 3 = 252

252 x 3 = 756

756 x 2 = 1,512

1,512 x 3 = 4,536

4,536 x 1 = 4,536

4,536 x 12 = 54,432

River Rapids

Multiply to find your way out of the river rapids.
If the path is correct, the answer to the first
multiplication sentence will be the first factor
in the next sentence, and so on.

Start **7**

Multiplication and Division Scholastic Professional Books

✳ $2.49 × 4 = ? ✳ 267 ÷ 5 = ? ✳ 742 × 54 = ? ✳ 43,097 ÷ 7...

Multiplying by 2- and 3-Digit Factors

Road Rally

The road to solid multiplication skills can take a lot of practice! With this multiplication game, students are well on their way.

◉→ Directions

1. Duplicate the *Road Rally* reproducible for each pair.

2. Discuss multiplying by 2-digit numbers, focusing on the regrouping process as follows:

Multiply by the ones. Regroup if necessary.	Multiply by the tens. Regroup if necessary.	Add the products.
	3	3
1	1	1
35	35	35
× 62 (2 × 35 = 70)	× 62 (60 × 35 = 2,100)	× 62
70	70	70
	2100	+ 2100
		2170

Ask students to give you a 3-digit number and a 2-digit number. Write a multiplication sentence on the board and let volunteers explain the multiplication process.

3. Review the terms *odd* and *even* and ask students to give examples of each.

4. Have partners play the game. You may want to let students check their answers with calculators.

✪ Taking It Farther

Talk about strategies that students used during the game. Then have them write four more numbers on the billboard and play a bonus round.

✓ Assessing Skills

✳ Do students complete the regrouping process from ones to tens and so on as necessary?

✳ Do they remember to regroup and add?

LEARNING OBJECTIVE

Students multiply by 2- and 3-digit numbers.

GROUPING

Pairs

MATERIALS

✳ *Road Rally* reproducible (p. 44)
✳ calculators (optional)

Road Rally

Play this game with a partner. Pick two numbers from the billboard and multiply them. Cross out the numbers and then find your score below. Play five rounds.

32	45	519	26	279
23	350	81	486	94
75	237	52	148	663
68	37	104	389	74

The product is even.
1 point

The product is odd.
2 points

The product is greater than 3,000.
3 points

Player 1

PRODUCT	POINTS
TOTAL	

Player 2

PRODUCT	POINTS
TOTAL	

What's New in the News?

Students "report" on numbers and practice multiplying.

⟳→ Directions

1. Ask students to look through newspapers and magazines and cut out two articles or ads that mention 2- and 3-digit numbers.

2. Divide the class into groups. Extra students will be fact checkers. The first student "reports" and reads the sentence that uses a number. The second student does the same. The third student multiplies, using the numbers as factors. A student checker verifies the product with a calculator. If the product is correct, the third student scores a point.

3. Group members change roles and continue until each person has a chance to multiply at least three times. You may also want to let checkers and group members exchange roles.

✪ Taking It Farther

Challenge students find 3-, 4-, or 5-digit numbers to use in a division game. The rules are the same as for the game above except that students divide these numbers by 2-digit numbers from articles or ads that they already have.

✓ Assessing Skills

Observe students who are having difficulty multiplying. Take time to review the multiplication process, using models if necessary. Encourage students to explain each step of the process.

LEARNING OBJECTIVE

Students multiply by 2- and 3-digit numbers.

GROUPING

Cooperative groups of 3

MATERIALS

* local and/or national newspapers
* magazines
* scissors
* calculators

Me First!

Almost everyone wants to go first! Students divide, order numbers and discover some interesting "firsts."

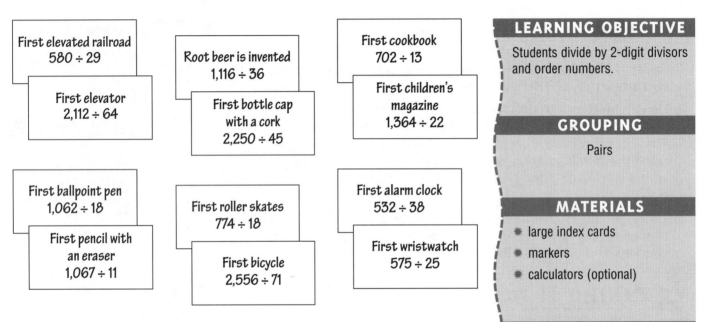

First elevated railroad
580 ÷ 29

First elevator
2,112 ÷ 64

Root beer is invented
1,116 ÷ 36

First bottle cap
with a cork
2,250 ÷ 45

First cookbook
702 ÷ 13

First children's
magazine
1,364 ÷ 22

First ballpoint pen
1,062 ÷ 18

First pencil with
an eraser
1,067 ÷ 11

First roller skates
774 ÷ 18

First bicycle
2,556 ÷ 71

First alarm clock
532 ÷ 38

First wristwatch
575 ÷ 25

Directions

1. Shuffle the cards before distributing to pairs.

2. Have students guess which fact on each card came first. Encourage them to use estimation.

3. Students take turns and divide. After they write the quotients, they order the numbers on each card. The number that is less, or "comes first," matches the event that came first.

Taking It Farther

Invite students to write two events in their own lives, with matching division problems to show which came first. Partners exchange cards, divide, and order the events.

Assessing Skills

You may want to have students use calculators to check their work. If they discover errors, ask them to explain each step of the division process.

LEARNING OBJECTIVE

Students divide by 2-digit divisors and order numbers.

GROUPING

Pairs

MATERIALS

✳ large index cards

✳ markers

✳ calculators (optional)

ADVANCE PREPARATION

Write the facts and division sentences on the front and back of the index cards as shown so each pair has a set of 6 cards. Each card will contain two different facts and sentences.

ANSWERS

Card 1— 20 (elevated railroad)
33 (elevator)

Card 2— 31 (root beer)
50 (bottle cap)

Card 3— 54 (cookbook)
62 (magazine)

Card 4— 59 (pen)
97 (pencil)

Card 5— 43 (skates)
36 (bicycle)

Card 6— 14 (clock)
23 (wristwatch)

Food for Thought

Students learn some fascinating food facts as they practice dividing by 2-digit numbers.

➔ Directions

1. Begin by reviewing odd and even numbers. Write several 3-digit numbers on the board and have students classify them as odd or even.

2. Write the following example on the board: 17,595 ÷ 85. Have students explain the division process, especially focusing on the zero in the quotient (207). Ask students to determine whether the quotient is odd or even and to explain how they know.

3. Duplicate and distribute a copy of the reproducible to each student or pair. Go over the directions and make sure they understand that if the quotient is odd, the fact is true.

☆ Taking It Farther

Ask students to develop rules that tell when quotients will be odd or even. Have them experiment with different combinations of odd and even dividends and divisors.

✔ Assessing Skills

Some students may forget to place a zero in the ones place of the second partial product. Provide a multiplication sentence and have them write the second factor in expanded form. Point out that the first factor is multiplied by the value of each digit in the second factor.

LEARNING OBJECTIVE

Students divide by 2-digit divisors.

GROUPING

Individual or pairs

MATERIALS

＊ *Food for Thought* reproducible (p. 48)

ANSWERS

1. 883, True

2. 482, False

3. 323, True

4. 895, True

5. 118, False

6. 623, True

7. 607, True

47

Food for Thought

These food facts may or may not be true! Divide and decide if the quotient is odd or even. If the quotient is odd, then oddly enough, the fact is true!

1. The shape of a pretzel was invented by a priest who gave "pretzels" to children when they memorized prayers.

   18,543 ÷ 21 = _____   **True**   **False**

2. Cracker Jack were the brainchild of a man named Jack Cracker, who invented the snack after some popcorn got stuck in his tooth.

   25,546 ÷ 53 = _____   **True**   **False**

3. Hot dogs can be traced back 3,500 year ago, when Babylonians stuffed animal intestines with spicy meat.

   19,703 ÷ 61 = _____   **True**   **False**

4. Pasta was first made in China from rice and bean flour.

   9,845 ÷ 11 = _____   **True**   **False**

5. Ketchup became popular when Alice White dropped a tomato and dipped some french fries in the juice.

   9,204 ÷ 78 = _____   **True**   **False**

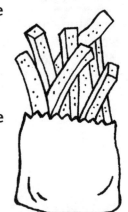

6. In 1902, American children received their first boxes of animal cookies. They were to hang the circus box with a string on their Christmas trees.

   28,035 ÷ 45 = _____   **True**   **False**

7. Frank Fleer's first try at creating bubble gum was called Blibber-Blubber Bubble Gum.

   43,097 ÷ 71 = _____   **True**   **False**

Sandwich by the Slice

No matter how you slice it, this activity will help students take a bite out of multiplying money.

⟳→ Directions

1. Copy and distribute the reproducible to students, and explain that they can ask 6 to 10 classmates to choose types of meat for their sandwiches. Each person can choose three slices of the same type of meat, or any combination of the three types of meat.

2. After compiling the data, students multiply to find the totals and the cost.

3. When students complete the tables, have them compare results.

✮ Taking It Farther

Ask students to estimate before they find the total cost, using front-end estimation and rounding. Discuss which method gave a more accurate estimate.

✓ Assessing Skills

Observe the steps that students use to solve the problems.

LEARNING OBJECTIVE

Students take a poll and multiply to solve problems.

GROUPING

Individual

MATERIALS

✳ *Sandwich by the Slice* reproducible (p. 50)

✳ calculators (optional)

Sandwich by the Slice

You're planning a party! You can invite from 6 to 10 classmates. You're going to order a sub sandwich from the Sandwich by the Slice Shop.

Each friend can order 3 slices of meat for the sandwich. Take a poll and find out what kinds of sandwiches your friends like. Then multiply to find out how many slices of meat the sandwich will have and the cost. (Don't forget to include yourself!)

Specials
Sandwich by the Slice
Turkey.......... 25¢
Roast Beef ...... 29¢
Ham........... 18¢

FRIEND'S NAME	HAM	ROAST BEEF	TURKEY

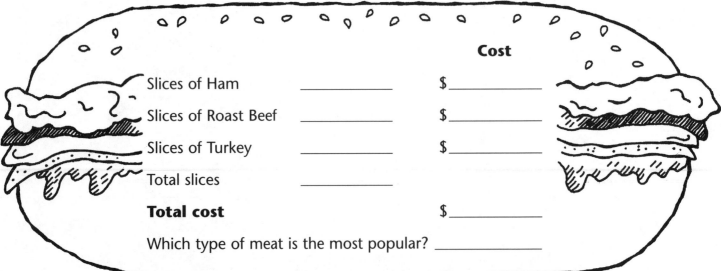

Cost

Slices of Ham _____ $_____

Slices of Roast Beef _____ $_____

Slices of Turkey _____ $_____

Total slices _____

Total cost $_____

Which type of meat is the most popular? _____

Shop 'Til You Drop

Students compare prices by buying in bulk and buying single items. Let them decide how to shop smart!

⟳→ Directions

1. Duplicate a copy of the reproducible for each student or group.

2. Students use newspaper ads to find the price of a single item and the price of the same item in a multi-pack. You may also want to have students research prices at the grocery store the next time their families go shopping, or supply prices yourself. Explore with them how to determine the amount saved. If necessary, help students find the savings by following either of these methods:

   a. Multiply the price of the single item by the total number of items in a multi-pack. Then subtract the cost of the multi-pack from that amount.

   b. Divide the price of the multi-pack by the number of items in it. Then subtract the cost of each multi-pack item from the price of a single item.

3. Discuss the results, including other items that students added to the table.

☆ Taking It Farther

Ask students to think of other types of items that are sold in bulk, such as socks, T-shirts, or barrettes. Encourage them to investigate savings and incorporate their findings in a newspaper or television ad for one of the products. Are multi-packs always the smarter buy?

✔ Assessing Skills

Determine whether students understand the computation to find the savings. If necessary, help them to analyze each step using a calculator for the computation.

LEARNING OBJECTIVE

Students develop consumer skills as they multiply and divide money amounts.

GROUPING

Individual or groups

MATERIALS

* *Shop 'Til You Drop* reproducible (p. 52)
* newspaper ads
* calculator (optional)

Shop 'Til You Drop

More and more people shop at warehouse-type stores where they can buy products in large quantities. Some grocery stores also sell items the same way.

How much do shoppers really save? Do some research to find out. Visit a store or look at newspaper ads to find some of the items listed below. Add some items of your own to the table. Then complete the table to find out if you're shopping smart.

ITEM	Cost of Single Item	Cost of Multi-pack	Number of Items in Multi-pack	SAVINGS
Paper Towels				
Canned Dog Food				
Bars of Soap				
Ballpoint Pens				
Canned Soda				

Party Problems

Students plan a party and multiply money amounts. The problem is figuring out the best way to spend $80!

⟲→ Directions

1. Duplicate a copy of the reproducible for each student or pair.

2. To begin, students make decisions regarding food, decorations, and other items that they would like to include in their party plan. Then they use information from local stores and restaurants to find appropriate prices. Students multiply as needed; they may also use calculators to determine the total cost.

3. If the total amount is more than $80, students should reassess their choices and make the appropriate changes. If the total is less than $80, they divide the cost by the number of students on the team to determine how much was spent for each student. Students round amounts to the penny as necessary.

4. Draw a blank check on the board and fill it in so students can refer to it when filling out their own checks. They may also draw additional checks on a separate sheet of paper.

5. Allow time for students to share and compare their results.

☆ Taking It Farther

Invite groups of students to plan a class party, spending as little money as possible. Challenge them to think about how they could make low-cost snacks and decorations.

✔ Assessing Skills

If students have difficulty multiplying money amounts, let them use bills and coins to review the regrouping process.

LEARNING OBJECTIVE

Students make decisions and multiply money amounts.

GROUPING

Individual or pairs

MATERIALS

✱ *Party Problems* reproducible (p. 54)

✱ paper

✱ markers

✱ calculators (optional)

Party Problems

You're planning the end of the season party for the basketball team. Seven players, including you, are coming. (Don't forget to invite the coach!) You can spend up to $80.

Use data from nearby stores and restaurants to make your plans. Decide what you will spend on decorations, food, and anything else you want to include. Then fill out the check from the coach! (If you need more than one check, you may draw them on a separate sheet of paper.)

Item	Cost	Number of Items	TOTAL

Paula Ramsey
34 Court Street
Oakton, IL

_____ 19 ___

$ _____

Pay to the order of _____

_____ Dollars

Paula Ramsey

Memo _____

How much did you spend on each player? _____

Multiplication and Division Scholastic Professional Books

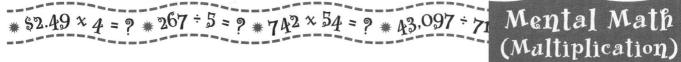

A-mazing Multiplication Patterns

Students look for multiplication patterns as they follow a maze to escape from a snake-infested desert.

Directions

1. Duplicate and distribute a copy of the reproducible to each student.

2. Write the following pattern on the board:

$$4 \times 2 = 8$$
$$4 \times 20 = 80$$
$$4 \times 200 = 800$$

Discuss the pattern and ask students to continue it.

3. Explain that students can look for patterns to help them through the maze on the reproducible. They begin with any multiplication examples at the left and find a path to the exit at the right. You may want to have students begin by finding the products for the examples on the rocks and looking for patterns before they begin the maze. If the path is correct, they will see a pattern. You may want to suggest that students make a list as they progress to verify the pattern.

4. An alternate approach is to have students complete the maze and then solve the examples to see if they form a pattern.

Taking It Farther

Have students create different multiplication patterns for the rocks on the path of the maze.

Assessing Skills

Observe whether students are able to use and verbalize their mental math strategies.

LEARNING OBJECTIVE

Students use mental math with multiplication patterns and properties.

GROUPING

Individual

MATERIALS

* *A-mazing Multiplication Patterns* reproducible (p. 56)

 Name _____ Date _____

A-mazing Multiplication Patterns

There aren't any gold or silver treasures in this desert—only snakes!

Enter the desert by any path on the left. To find your way out as quickly as possible, look for a multiplication pattern on the rocks.

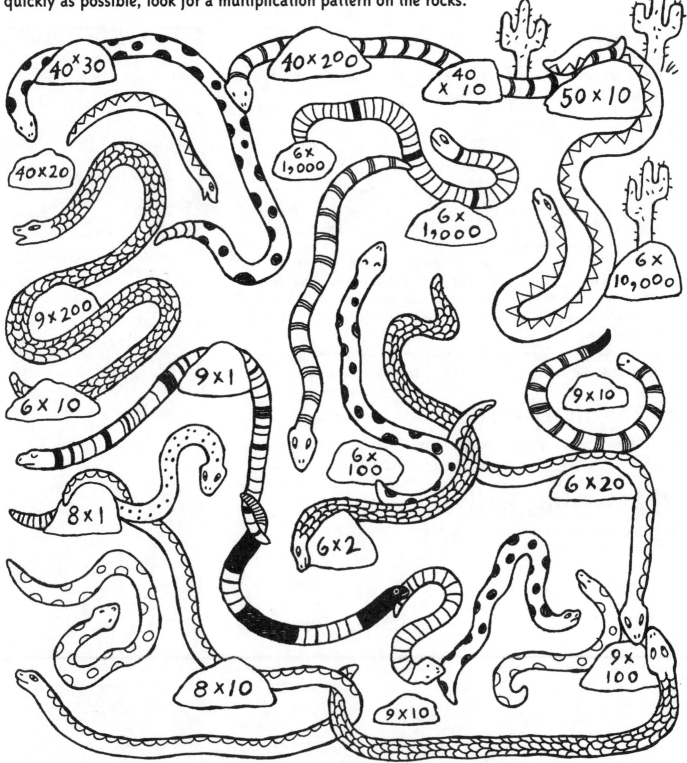

Multiplication and Division Scholastic Professional Books

Treasure Hunt

Will students fall into Luckless Lagoon, or will they find the treasure? It all depends on their mental math skills!

◌→ Directions

1. Write 15,000 ÷ 3 on the board and discuss how to use mental math to divide. Talk about using patterns and attaching the correct number of zeros.

2. Pairs write 10 division problems such as 6,400 ÷ 8, where the dividend is a multiple of 10, 100, or 1,000 and the divisor is a 1-digit number. Each pair exchanges a set of cards with another pair of students.

3. Distribute the reproducibles and explain how the game is played. The pairs make a stack of the cards and use the counters as markers. The first player chooses a card and uses mental math to find the quotient. The partner uses a calculator to check.

4. If the quotient is correct, the player rolls the number cube and moves that number of squares on the map. The player may move horizontally or vertically. If the player lands on Brigadoon Island, Craggy Caverns, or Luckless Lagoon, he or she loses a turn. If he or she lands on Frightful Falls, the player "slides" down.

5. Players need an exact number to reach the treasure at Finish.

☆ Taking It Farther

Let students make more cards with different numbers and play the game again. Challenge them to include 5-digit multiples.

✓ Assessing Skills

Ask students to describe the relationship between the number of zeros in the dividend and in the divisor.

LEARNING OBJECTIVE

Students divide using mental math.

GROUPING

Pairs

MATERIALS

For each pair:

✳ *Treasure Hunt* reproducible (p. 58)

✳ 10 index cards

✳ 2 counters

✳ markers

✳ calculator

✳ number die

Treasure Hunt

Ahoy there, mates! Are ye searching for buried treasure?
It's not a matter of luck—it's all in your mind!

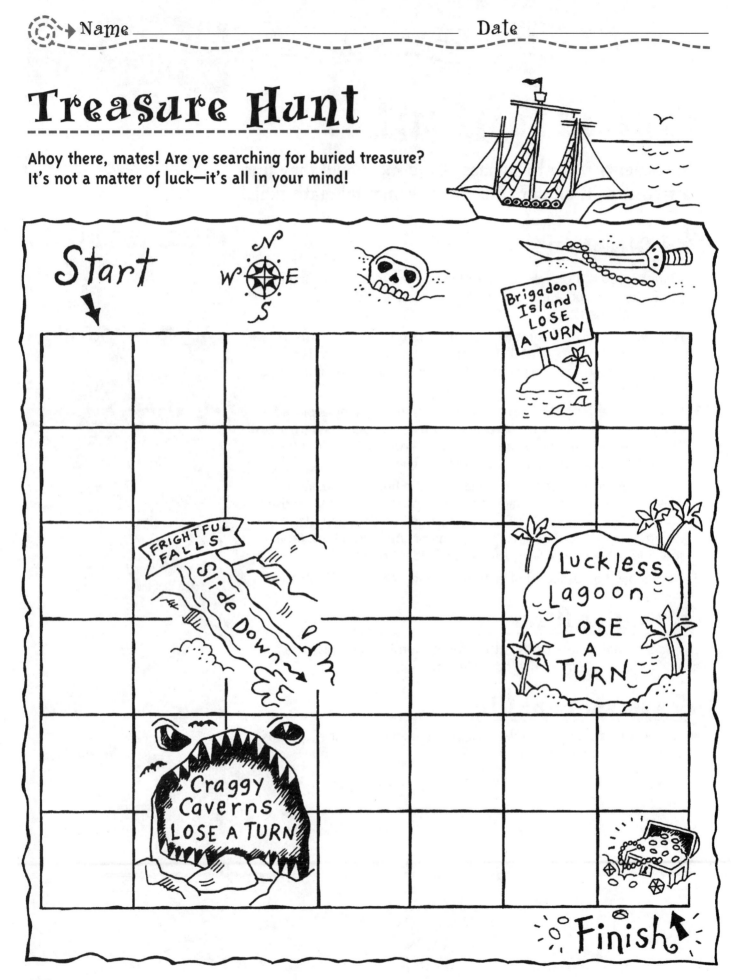

Start

N
W E
S

Brigadoon Island LOSE A TURN

FRIGHTFUL FALLS

Slide Down

Luckless Lagoon LOSE A TURN

Craggy Caverns LOSE A TURN

Finish

Multiplication and Division Scholastic Professional Books

Do I Have Problems!

Use these quick skill builders as a class warm-up, as a time filler for students who finish a test early, or just for a fun break from the textbook!

CROSSNUMBER PUZZLE

Multiply and divide to solve the puzzle.

a.	b.		c.	d.	
e.		f.		g.	h.
		i.	j.		
k.	l.		m.	n.	
	o.	p.			
		q.			

ACROSS

a. $72 \div 3 = $ [24]
c. $5 \times 13 = $ [65]
e. $12 \times 24 = $ [288]
g. $9 \times 8 = $ [72]
i. $3 \times 8 = $ [24]
k. $490 \div 14 = $ [35]
m. $8 \times 31 = $ [248]
o. $9 \times 5 = $ [45]
q. $1,000 \div 10 = $ [100]

DOWN

a. $11 \times 2 = $ [22]
b. $3,120 \div 65 = $ [48]
d. $3 \times 19 = $ [57]
f. $738 \div 9 = $ [82]
h. $4 \times 52 = $ [208]
j. $6 \times 7 = $ [42]
l. $9 \times 6 = $ [54]
n. $8 \times 55 = $ [440]
p. $1,020 \div 20 = $ [51]

BODY OF BONES

How many bones are in the human body? Find the products and then the sum. Write the .

Skull	$1 \times 29 = $ ____	[29]
Spine	$2 \times 13 = $ ____	[26]
Chest	$5 \times 5 = $ ____	[25]
Hands	$2 \times 27 = $ ____	[54]
Arms	$2 \times 5 = $ ____	[10]
Legs and Feet	$2 \times 31 = $ ____	[62]
Total		[206]

There are ____ bones in the body! [206]

PUZZLING POUNDS

Petra has 126 pounds of apples to sell. She has 39 bags in all. Some are 2-pound bags and some are 5-pound bags. How many of each does Petra have?

[23 2-pound bags and 16 5-pound bags]

BOXED IN

Find the numbers that complete each multiplication box.

×	2	7		5	8
4	8				
1			3		
	0				
5				25	
6					

[Row 1—3; Row 2—28, 12, 20, 32; Row 3—2, 7, 5, 8; Row 4—0, 0, 0, 0, 0; Row 5—10, 35, 15, 40; Row 6—12, 42, 18, 30, 48]

CALCULATED GUESS

Use different combinations of the given numbers to complete each multiplication sentence. Use a calculator to help you find the missing numbers.

1   3   5   8

__ __ × __ __ =   630   [18 × 35]
__ __ × __ __ =   1,245   [15 × 83]
__ __ × __ __ =   1,938   [51 × 38]
__ __ × __ __ =   2,835   [35 × 81]

SHAPE UP

Each shape represents a number. Find the missing factor or divisor and write it inside in each shape.
[triangle = 3, square = 8, hexagon = 16, circle = 4]

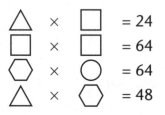

△ × □ = 24
□ × □ = 64
⬡ × ◯ = 64
△ × ⬡ = 48

MATCH GAME

Find each product and dividend. Draw a line to match them.

96 × 62 = [5,952]   [5,808] ÷ 12 = 484
72 × 84 = [6,048]   [5,952] ÷ 12 = 496
24 × 242 = [5,808]   [6,048] ÷ 36 = 168

GREATEST POSSIBLE PRODUCT

Find the number of digits in the greatest possible product of a 2-digit factor and a 3-digit factor. Write the factors and the product.

__ __ × __ __ __ = _____ [99 × 999 = 98,901]

How many digits? _____ [5]

MULTIPLICATION MAZE

Draw a line connecting correct multiplication sentences from the entry arrow to the exit arrow of the maze. [14 × 86 = 1,204; 38 × 17 = 646; 78 × 124 = 9,672]

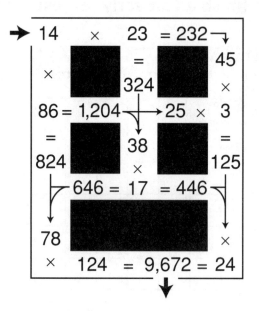

PYRAMID PRODUCTS

Each number in the pyramid is the product of the two numbers below it. Complete the pyramid.

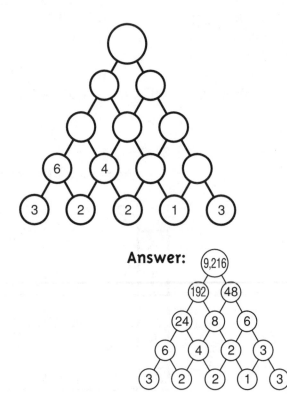

Answer:

9,216
192   48
24   8   6
6   4   2   3
3   2   2   1   3

60

FIND THE KEYS

Each key unlocks a door that is marked by factors that match the product on the key. Each box represents a room. To mark the path out of the house, shade the boxes that have factors that match a key. There are no diagonal moves. [The following rooms should be shaded—576 × 1; 36 × 16; 24 × 24; 52 × 36; 105 × 10; 144 × 13; 36 × 16; 48 × 12; 25 × 42; 18 × 32; 78 × 24.]

Keys: 576, 1,872, 1,050

ENTER

576 × 1	36 × 16	24 × 24	52 × 36	105 × 10
26 × 72	150 × 7	21 × 50	23 × 41	144 × 13
34 × 18	250 × 5	36 × 24	48 × 12	36 × 16
83 × 26	144 × 12	18 × 32	25 × 42	48 × 52
55 × 11	17 × 82	78 × 24	150 × 4	62 × 31

LEAVE

PICNIC PLANNING

Dan's Deli Menu
Tuna $3.50
Turkey. . . $4.25
Egg. $3.25

a. Alyssa is buying sandwiches from Dan's Deli for a class picnic. She has orders for 8 tuna, 12 turkey, and 5 egg sandwiches. How much money will she need to cover the cost of sandwiches? (Dan is not charging her tax.) [$95.25]

b. Dan sold $112.00 worth of tuna sandwiches, $174.25 worth of turkey sandwiches, and $65.00 worth of egg sandwiches on Tuesday. How many of each kind of sandwich did Dan sell? [32 tuna, 41 turkey, and 20 egg]

SIX-FACED CUBE

* Write a multiplication sentence to describe how many squares are on each face of the cube. [3 × 3 = 9]

* Write a multiplication sentence to describe how many squares there are on all faces of the cube. [6 × 9 = 54]

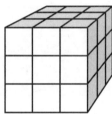

TARGET FACTORS

Circle the two numbers on each target that have a product closest to the score.

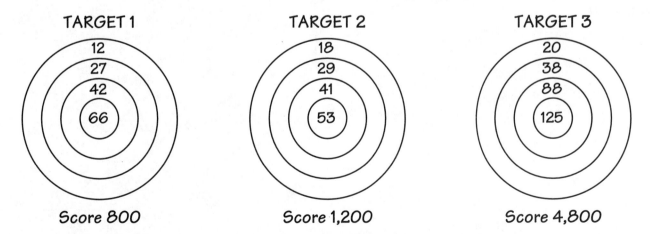

TARGET 1
12
27
42
66
Score 800

TARGET 2
18
29
41
53
Score 1,200

TARGET 3
20
38
88
125
Score 4,800

[Target 1: 12 × 66 = 792; Target 2: 29 × 41 = 1,189; Target 3: 38 × 125 = 4,750]

MULTIPLICATION COUNT

Use multiplication to find the number of squares in each section in the diagram below. Write and solve a multiplication sentence for each section.

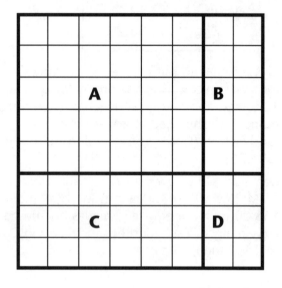

A. _____

B. _____

C. _____

D. _____

✳ Add to find the total number of squares in the diagram.

✳ Write a multiplication sentence to find the total number of squares in the diagram.

[A—5 × 6 = 30; B—5 × 2 = 10; C—3 × 6 = 18; D—3 × 2 = 6; total number = 64 squares; 8 × 8 = 64]

MULTIPLICATION WHEEL

Complete the multiplication wheel by finding each product.

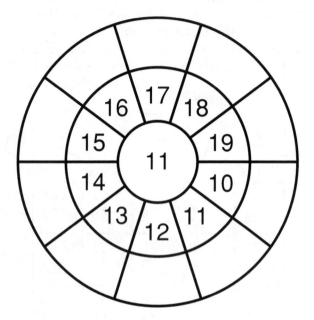

[11 × 10 = 110; 11 × 11 = 121; 11 × 12 = 132; 11 × 13 = 143; 11 × 14 = 154; 11 × 15 = 165; 11 × 16 = 176; 11 × 17 = 187; 11 × 18 = 198; 11 × 19 = 209]

Name ____________________ Date ____________

STUDENT SELF-EVALUATION

In My Opinion

The activity ______________________________ was:

Easy Hard

because:

____

____

____

____

My work on this activity was:

poor fair good excellent

because:

____

____

____

I used the following math strategy or strategies:

- ____________________
- ____________________
- ____________________
- ____________________
- ____________________
- ____________________

I would share this tip with someone who is about to do this activity:

____

____

____

TEACHER ASSESSMENT FORM

Student						
UNDERSTANDING						
Identifies the problem or task.						
Understands the math concept.						
SOLVING						
Develops and carries out a plan.						
Uses strategies, models, and tools effectively.						
DECIDING						
Is able to convey reasoning behind decision making.						
Understands why approach did or didn't work.						
LEARNING						
Comments on solution.						
Connects solution to other math or real-world applications.						
Makes general rule about solution or extends it to a more complicated problem.						
COMMUNICATING						
Understands and uses mathematical language effectively.						
COLLABORATING						
Participates by sharing ideas with partner or group members.						
Listens to partner or other group members.						
ACCOMPLISHING						
Shows progress in problem solving.						
Undertakes difficult tasks and perseveres in solving them.						
Is confident of mathematical abilities.						

SCORING RUBRIC

3	2	1
Fully accomplishes the task.	Partially accomplishes the task.	Does not accomplish the task.
Shows full understanding of key mathematical idea(s).	Shows partial understanding of key mathematical idea(s).	Shows little or no grasp of key mathematical idea(s).
Communicates thinking clearly using oral explanation or written, symbolic, or visual means.	Oral or written explanation partially communicates thinking but is incomplete, misdirected, or not clearly presented.	Recorded work or oral explanation is fragmented and not understandable.